AMAZON ECHO

DOT

I0492695

The Ultimate Guide for Complete
Beginners On How to Setup Your Amazon
Echo Dot in Few Minutes.

BY

CHARLES S. MILLS

COPYRIGHT

Charles S. Mills

TABLE OF CONTENT

CHAPTER 1

INTRODUCTION

The Amazon Echo Dot is a handy device that can both be seen and heard as a smart home device. As an amazing device it has gained popularity through these years, with over 3 million users who owns it, and enjoys using the amazing features it has in it making it interesting and comforting to the fullest.

Apart from the popular use of Amazon Echo Dot device as an E-reader, it can also perform other functions you can think of including weather forecast, news update, alarm, and more. Everything you have ever dying to know about your Echo Dot including what is the device, how to control a smart home, detailing what skills it has, troubleshooting common problems, setup the connected device, and much more. The user interface is a friendly kind and it has a collection of amazing stuff in it.

This guide will show you how you can setup your Amazon Echo Dot, all you have to do to achieve and master the Amazon Echo Dot device is carefully follow it step by step as they are instructed on this guide. Also it can perform so many things when you master and unleash it full potentials, but if you find any issue very difficult, there is no needs to worry because this guide got u cover.

Thankfully each steps are very easy and simple to understand with proper meditations, making it more interesting and enjoyable. That even a beginner can master it in a few minutes and unleash its full potential.

CHAPTER 2

AMAZON ECHO DOT

The Echo Dot is the entrance to a world of smart home automation and fun with your virtual assistant. Meanwhile, Amazon offers several Echo devices, the Echo Dot is a very good choice due to its minimum tag and thin profile.

During the setup of your Echo Dot, if you are stuck up or need assistant finding out the basics, you just arrive at the right place. In this guide, everything you've ever trying to figure out to get started and acquitted with your Echo Dot and utilize its power to its fullest.

Note: Amazon renew the Echo Dot in fall 2016. The two generations have few similarities. Therefore, this guide is cover with the second generation model, we will touch some few places where they both have their differences.

Below is the rundown of everything you can expect from this great guide:

a. Your Echo Dot setup

b. Basic Alexa commands and additional skills

c. Important Echo Dot functions

d. Alexa app settings adjustment

e. Troubleshooting your Echo Dot

CHAPTER 3

UNBOXING AND SETTING UP THE ECHO DOT

Before getting started, you need to remove the Echo Dot from the box it came with. Inside it, you should find the following items:

→ A power adapter to enable you plug into the wall.

→ Items to try card with some sample Alexa commands.

→ The Echo Dot unit – we are going to refer it as the Echo or Dot from here.

→ A standard micro USB cable to power the unit.

→ Starters pamphlet with important setup instructions.

Now, to get started first thing to do is plugging the micro USB cable into the back of your Echo Dot. Then plug the standard USB end into the adapter, then into a wall socket. You should place your Echo Dot in a central position in a room so it can hear your command from anywhere. Its microphones are solid and sensitive; therefore, you should avoid playing around with it.

When your Echo Show startup it will show a blue light. Then give it a few minutes to finish its initialization process. Immediately you see an orange ring of light at the top, Alexa will say you are ready to get online now.

HOLD THE ALEXA APP

As long the Echo Dot doesn't have a screen, the setup process will continue on your mobile device. Ensure to install the Alexa from the right app store.

→ Amazon Alexa on Android.

→ Amazon Alexa on iOS.

→ Failing to have a smart phone you can use the Alexa web portal.

Next, you should open the Alexa app, and then sign into your Amazon account (or you can create one if you don't have). The Amazon app might automatically recognize your account if you have already use it on your mobile device.

amazon alexa

Sign In   Forgot password?

Email (phone for mobile accounts)

Amazon password

☐ Show password

☐ Keep me signed in. Details

SIGN IN

New to Amazon?

CREATE A NEW AMAZON ACCOUNT

Immediately you are done signing in and accept the terms and condition of use, a list of Echo devices will appear. Since you are setting up an Echo Dot, so you should choose that option. Then confirm your language option, next tap on the **Connect to Wi-Fi** button. Since your device in still plugged in, the light ring will remain orange as it advises. Then hit the **Continue** button.

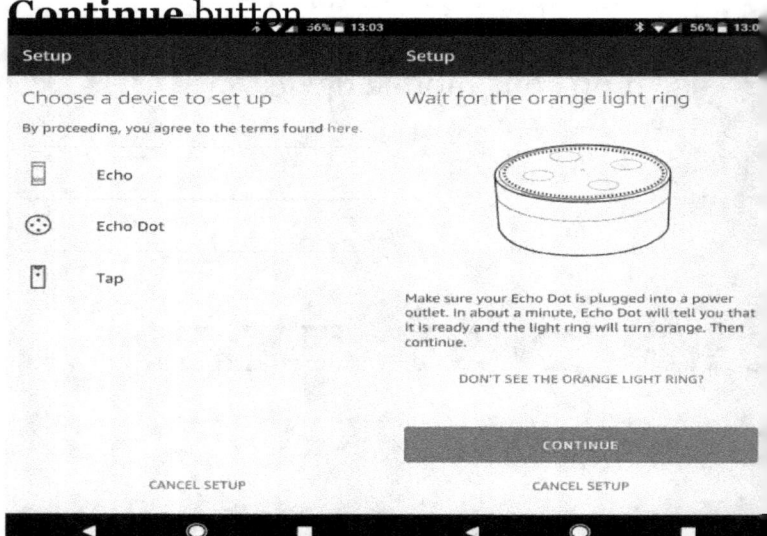

Automatically your phone will then try to connect to your Echo Dot. But if this doesn't solve the issue, the app will then request that you should hit and hold the

Dot's action button (the one with a bump) for a few seconds. Immediately the device is found, ensure to hit the **Continue** button again.

After you are done with that, you will need to add the Echo to your Wi-Fi network. Type your network name here, then type the password. Your Echo will go online a moment after you click on **Connect.**

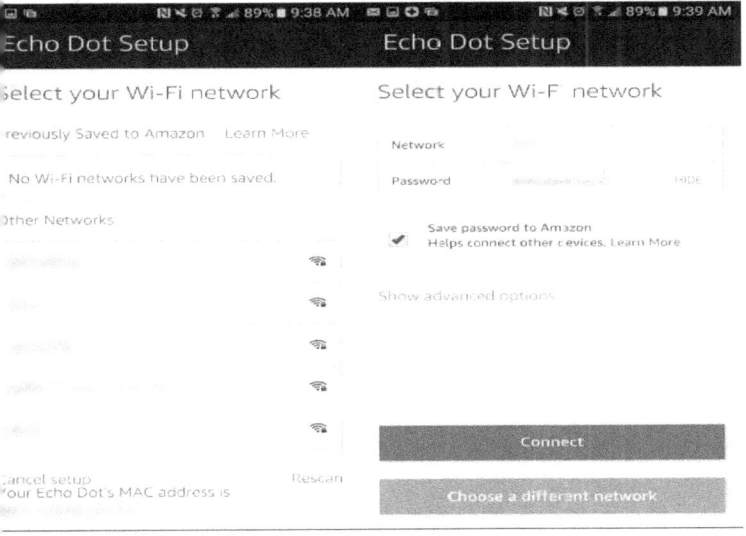

Finally, the last step is you will need to decide on how to hear your Echo.

However, there are three options for this: **Audio Cable, Bluetooth,** and **No Speakers.** The Dots permit you to connect your device to a speaker by making use of Bluetooth or an audio cable for better audio. However, if you decide not to use either of these, the final option will play all audio through the Dot's basic speakers.

Echo Dot Setup

Select how you want to use your Echo Dot

Bluetooth
Pair Echo Dot with your speaker using Bluetooth.

Audio Cable
Connect Echo Dot to your speaker using an audio cable.

No speakers
Use Echo Dot as a stand-alone device.

For now, you should select **No speakers** and then we'll discuss the other options later on this guide.

Once you have completed the setup, the app will offer to show you a quick video using the Alexa, and it gives a few sample commands at you. If need be, this can be review; we are going to discuss using the Alexa below.

CHAPTR 4

IMPORTANT ALEXA SKILLS AND COMMANDS

Now that the Echo is ready, you can communicate to it any time just by saying the wake word **Alexa** next by a command. For example, say **Alexa, what date is it?** And your Echo will tell you. We will cover some commands, but don't worry to give it a try. At worst, Alexa will reply you back that she doesn't know.

BUILT-IN ALEXA COMMANDS

Your Echo Dot has the ability to do lot right out of the box. Below is the sampling.

Alexa...

a. **Give me my Flash Briefing:** Your Echo Dot will give you the latest news headlines and update. You can as well customize your sources, which we are going to discuss later.

b. **What date is it?** Alexa can help in case you ever wake up after being frozen for very long time or just forget what day it is.

c. **Switch on the lights.** Alexa can behave as a one-stop hub for controlling your house just with some good smart home devices.

d. **Set alarm for 7am.** You can set an alarm on your Echo without using your clock or phone's buttons. Also, a schedule can be setup by simply saying **Set a repeating alarm for 7am weekdays.**

e. **Track my order.** If need be, you wish to know when your new Amazon package is going to arrive? Alexa will tell you.

f. **Set a timer for 5 minutes.** This works perfectly well when you are cooking something with dirty hands and you wish to wash them.

g. **Play the Kenny G station on Pandora.** on this, more details will be given on music below.

h. **Order for laundry detergent.** Your Echo allows you order from Amazon simply using your voice. Say the name of the goods you wish to purchase (failure to be specific enough, Alexa will list popular options). But in case you are not sure, you can send goods to your Amazon cart instead.

i. **How is the traffic?** if your office address is set in the settings (see below), Alexa will give you a correct detail about your commute time.

j. **What restaurants are nearby?** Locate the nearest place to relax and have a bite to eat.

k. **Stop.** You can use this global command to end audio playback, or you can just shut Alexa up if she's talking too long.

l. **What is the extended forecast for McMurdo Station Antarctica?** If you ask Echo about the weather forecast without being specific of the place, your Echo will answer you with your current location.

m. **Add finish building my PC on my to-do list.** Your Echo can create a list of tasks to lessen the stress using Alexa.

n. **Add books to my shopping list.** Here, Echo can as well make a shopping list making it easier to recall what you need to purchase.

In most command, additional information will be seen in the Alexa app. Hit the **Home** button at the bottom of the screen to have a view of everything you have recently commanded your Echo. For example, after asking Alexa about nearby Chinese restaurants, she provided a few options while scrolling the app enable you to have reviews, business hour, and addresses.

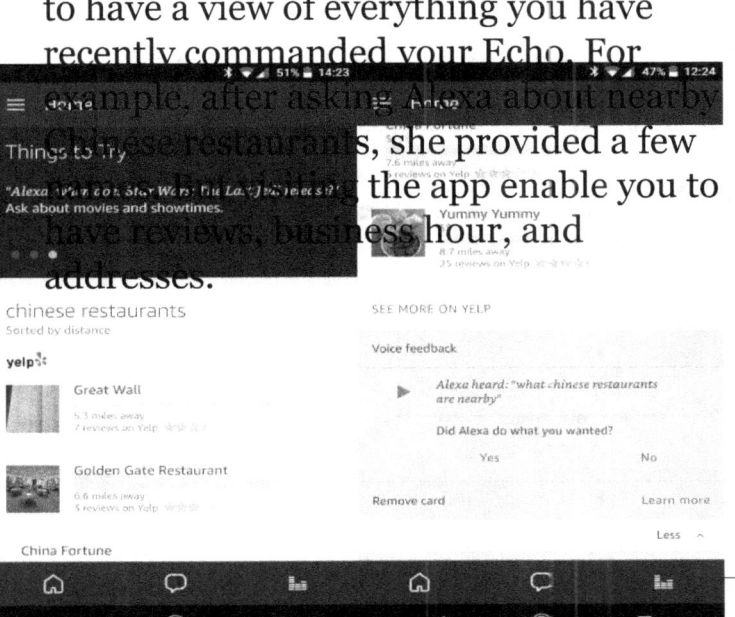

When searching for anything new to test Alexa, there is a whole section of the app to search. On the left menu, you can slide open and then select **Things to Try** for a list of options.

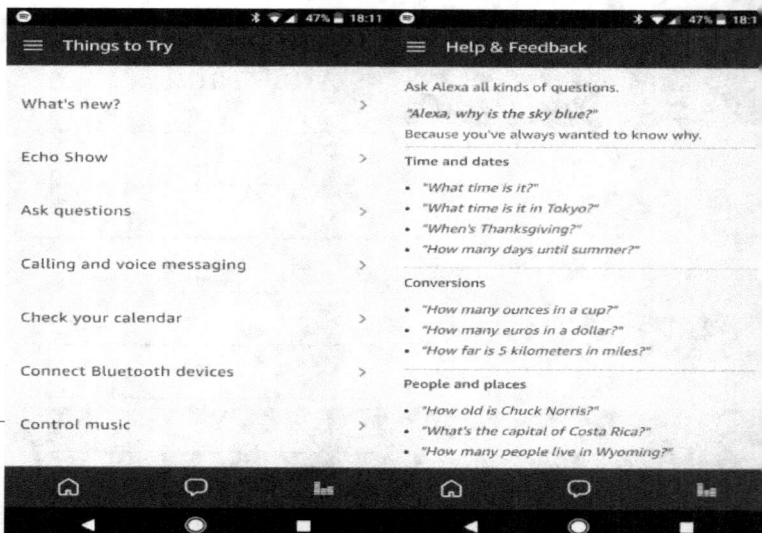

AMAZON ECHO DOT SKILLS

Much appreciations to the talented developers, there isn't any limit for you to the default Alexa skills. Anyways, to expand the functionality of your Echo Dot, there is a world of skills you can browse and add to it. To have a look at them, open the Alexa app and select the **Skills** tab from the left menu.

You will find the skills storefront. There is a lots happening there, but searching for new skills isn't difficult. You can browse through the front page and you will find a list of some popular options, along with the latest additions. Scroll down to the bottom and you will find some categories like **Local, Productivity,** and **Health**

and Fitness. You can also look for a skill using the top bar.

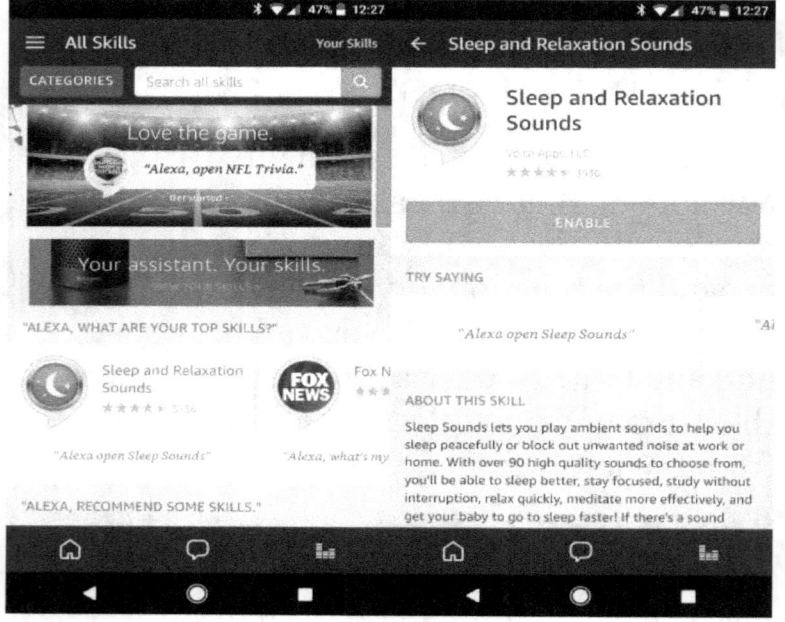

You should select a skill to read more about it. Each one gives the phrases you need to ask Alexa to enable open the skill. If need be, you can also go through the developer's description and see reviews.

All you have to do is just hit the **Enable Skill** button at the top to add into your Echo. After some time, you can command Alexa to launch it. You can command Alexa to add skills as well. Meanwhile, that isn't as useful as you are not able to browse a visual list.

We have written all basic Alexa skills and some of the amazing skills too, so you can refer to those list for some skills to check out.

CHAPTER 5

VITAL ECHO DOT FUNCTIONS

Since you have completed the setup, you can ask Alexa some questions. But to have a full part, you should be enlightened about some of the other functions of the Echo Dot.

ECHO DOT LIGHTS & BUTTONS

We haven't talk about the buttons on your Echo Dot unit yet, now let's have a look at each button at the top.

a. **Microphone off Button:** To disable the Echo Dot microphones, you should hit it. The Echo will show red color to tell you it's disabled and won't be able to respond to wake word. To enable it again, you can press it.

b. **Action Button:** This is the button with a dot at the top of the Echo. Press it to start up your Echo just like saying the wake word. And tapping it also ends a ringer alarm or timer.

c. **Plus, and Minus Button:** This control the volume when you hit one. You will see that the white light ring around the Echo Dot bibs or grows to show the current volume. To set a prefer volume level, you can also say **Alexa, volume five,** any digit between 1 and 10 inclusive will work.

Note that possessing a first generation Echo Dot, you can control the volume by rolling the outside ring. The earlier model doesn't possess a volume buttons.

Your Echo Dot will sometimes light up with different colors and patterns to talk with you. You should be acquitted with these common ones:

a. **Solid Red:** To disable the microphone by using the button.

b. **Solid Blue with Spinning Cyan Lights:** This appear when the device is starting up. But if you find this regularly, you might be accidentally unplugging your device.

c. **Pulsing Green Light:** This appear when you received a message or call.

d. **Solid Blue with Cyan Sliver:** This appears when the Echo Dot is processing or checking out what you said.

e. **Waves of Violet:** This appears when setting up Wi-Fi, the device encounters an error. Refer to the troubleshooting section below on this guide if you see this much often.

f. **Pulsing Yellow Light:** It appears when you receive a message. You can say **Alexa, play my message** to listen to it.

g. **All Lights off:** It appears when the Echo is in standby mode and is listening for your commands.

h. **Flash of Purple Light:** This appears when Alexa processes a command. And it means that your device is in Do Not Disturb mode.

INCLUDING MUSIC ACCOUNTS

Of all functions, one of the most useful of an Echo Dot is playing music. Giving the Echo Dot just one command, you can set the mood for an event or get your best music playing. It's a lot faster than scrolling through the menus on your mobile device for the suitable tunes.

Unless you are streaming music through your Amazon account, to access your libraries you will need to connect your accounts. To do this, you should open the Alexa app on your mobile device and slide out the left menu. Select **Settings,** then move down and tap **Music and Media.**

once you are here, you will find available music services including Amazon Music, iHeartRadio, Pandora, and Spotify. Then you should hit the link next to the service you want to connect and follow the procedures to sign in and again link it to your Echo Dot.

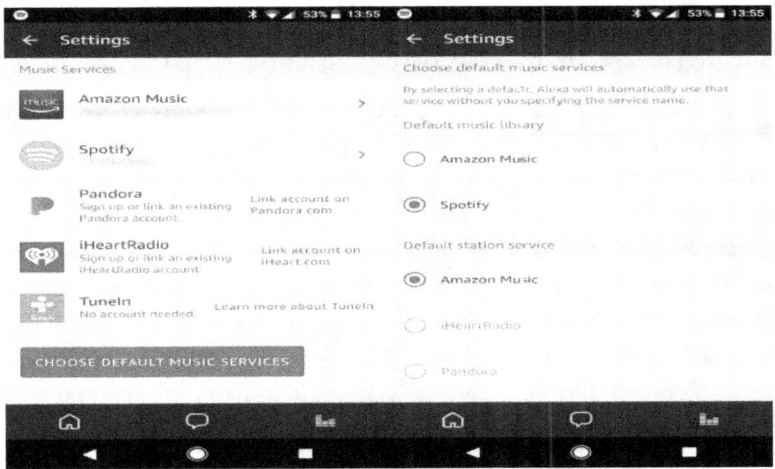

At the moment you are signed in, it's worth hitting the **Choose Default Music Service** button. This allows you indicate what service is primary. For example, if you give a command to Alexa like, **Alexa, play some jazz music** and use Spotify sets as default, Alexa will always play from Spotify. But if you want to use a different service as your primary, you will need to add **"from Spotify"** anytime you request to play music.

VIDEO AND BOOKS

Move out of the left menu and hit the **Music, Video, and Books** option, and you will find lots of services. Remember we have already talked about music, but Alexa still got a few more tricks in her.

Beneath the **Video** section, your Echo can be connected to a Fire TV or Dish Hopper Smart DVR. These allows you to control playback using your voice, which is pretty slick.

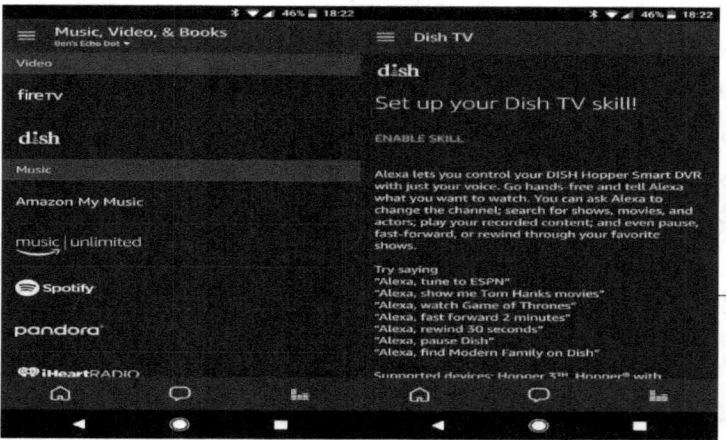

If you are more of a bookworm, move down and you can access your Audible and Kindle libraries on your Echo. Alexa can read your Audible audiobooks as well as every Kindle eBooks you have. They must come from the Kindle store, and Alexa can't read comics or graphic novels. You also won't be able to reduce the reading speed. Still, it's a suitable way to meet up on reading even while you are busy working.

GET YOUR ECHO DOT CONNECT TO EXTERNAL SPEAKERS

Echo Dot contains good speakers that works perfectly well for talking to Alexa. While they won't fill the room with sound, they get the job done for basic commands. Meanwhile, they are noticeably sub-par when it comes to play music. However, you might want to get your Dot connected to a speaker via Bluetooth or an audio cable for good sound quality.

Wanting to use a 3.5mm audio cable, just simply plug one end of the cable into your Echo and then the other end plugs it into the speakers you want to make use of. Hence, whenever Alexa plays any audio, through the speakers you will hear it. Ensure to put the volume level of both devices at a suitable level if you can't hear it loud and clear. Note that you won't be able to do reverse and use the cable to play music on your Dot from another device.

Trying to connect a Bluetooth speaker, just take a few taps under the settings menu. Open the Alexa app, slide out the left menu, and choose **Settings.** Here, at the top of the list, you should find your device under the **Devices** heading. Choose it, then you should select **Bluetooth.**

Gotten the resulting menu, press the **Pair a New Device** button and place your speakers in pairing mode. Choose the right device and your Echo will play all audio through that same speaker. If you wish to disable the connection, just give a command to Alexa by saying **Alexa, disconnect Bluetooth** and the audio will continue to play through your Dot. If you want to reconnect on a later time, just say **Alexa, connect Bluetooth.** Ensure to put your speaker on when doing so.

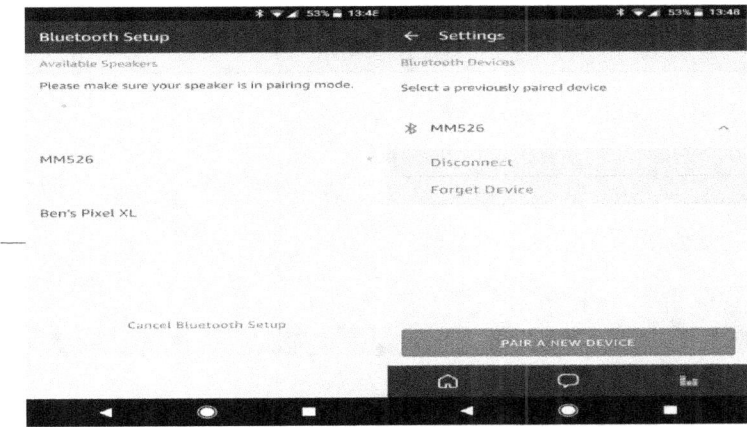

VOICE CALLING AND MESSAGING

Free calling and messaging to other Echo devices is one of the latest features. You can use this feature to drop a message for a friend, or even call them and talk live at no extra cost. Hit the **Messaging** tab at the bottom of the app screen to see your recent messages and even send one from the app.

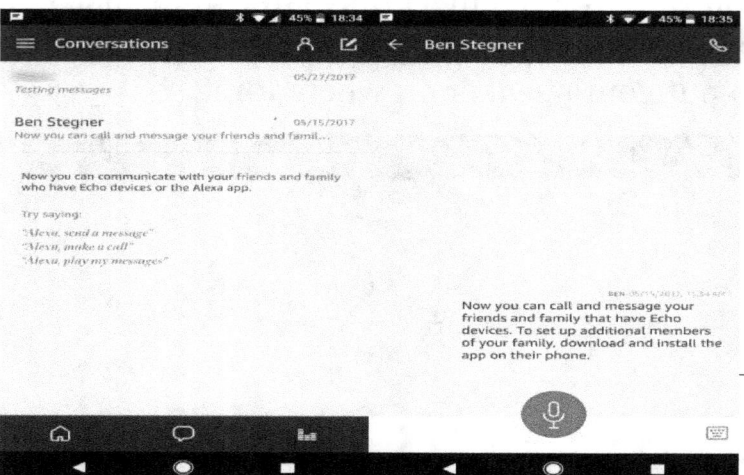

You can refer to our guide for Echo calling for more info.

THE SMART HOME FUNCTIONALITY

The Echo Dot is a good cornerstone to a smart home setup. You should slide to open the menu and choose the **Smart Home** option, and you will find the Echo's hub for adding new devices and tweaking them. Rushing into creating a smart home is far way beyond the scope of this article, have a look at our smart gadgets that are simple to setup and our $400 smart home starter kit if you want to tap into this side of your Echo.

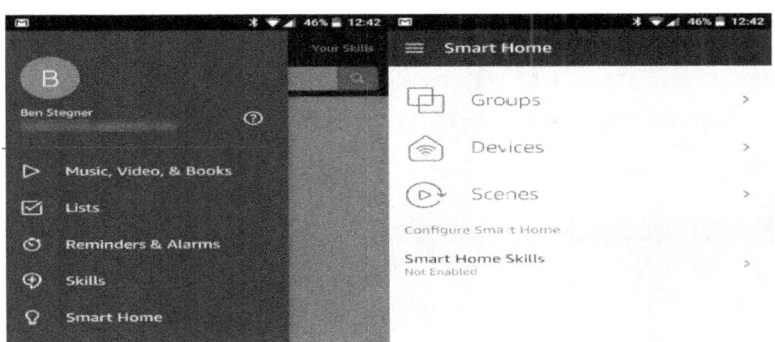

CHAPTER 6

SETTINGS OF AMAZON ALEXA APP

We have discussed about the basic using your Echo Dot, along with using Alexa and adding skills. As someone who owns an Echo, you should be in line with the useful options hanging out in the Alexa app on your mobile device. To have access to it, you should just slide out the left menu and choose **Settings**. Now, let's outline what you can perform in there.

CHANGING THE WAKE WORD OF YOUR ECHO DOT

Normally, the wake word for all Echo device by default is **Alexa.** But it can be change to any other names, especially when someone living in your house bears a name similar to it. Head to **Settings,** click on your device's name, and then move down to **Wake Word.** You can select from four options: **Alexa, Echo, Computer,** and **Amazon.**

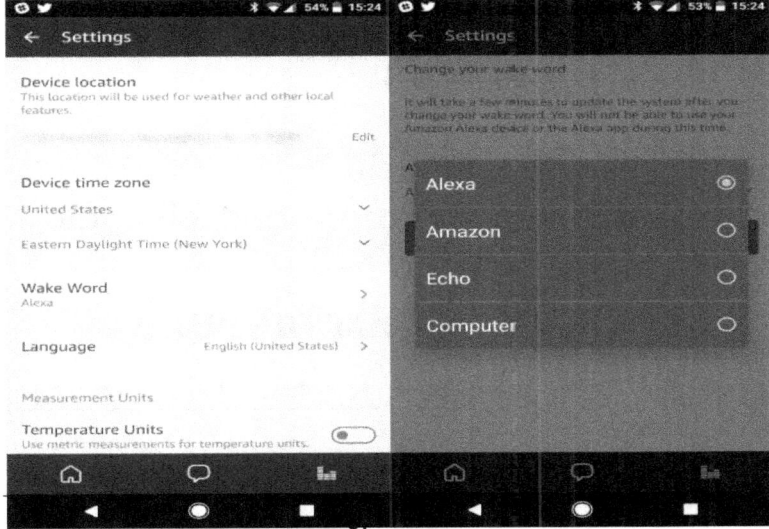

The last one should appeal to Star Trek fans. But even if you wouldn't mind **Alexa,** by means of changing your wake word can protect your device from getting hijacked.

PAIRING A REMOTE WITH YOUR ECHO DOT

Keeping costs down is not added in the box, but Amazon provided a remote for the Echo Dot. If you buy one, go to:

→ **Settings**

→ **Device**

→ **Pair device remote,** and follow the steps to sync it.

ECHO DOT DO NOT DISTURB TURN ON

Having lots of friends that also owns an Echo device, you probably won't want to be disturb by them sending messages all the time. Move into **Settings**

Device, and then switch on

Do Not Disturb, and Alexa won't inform you if anyone call or text you. You can also schedule a suitable time for Do Not Disturb to activate every day automatically.

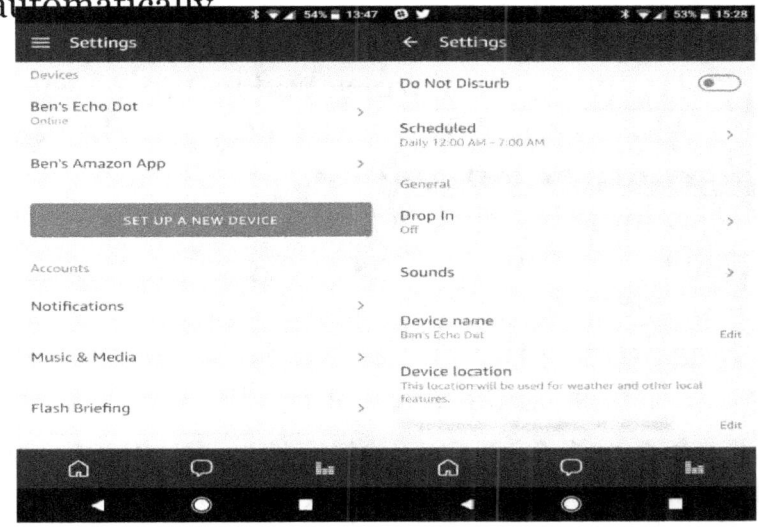

To do this, just say **Alexa, switch on/off Do**

Not Disturb to get this change without digging in the menus.

REDUCE ECHO DOT SOUND OPTIONS

Having issues with your Echo Dot doesn't play super sounds very much, you can change the options it offers to your taste. Head to **Settings** and type your device's name, then tap **Sounds.**

Try out a new alarm sound by hitting the **Alarm** entry. Ensure to put your **Alarm, Notification Volume, and Timer** in a volume high enough so you are able to hear it. You wouldn't like to miss an alarm! But you can also turn off the **Audio** option under **Notifications** if you are busy and don't want to be inform when

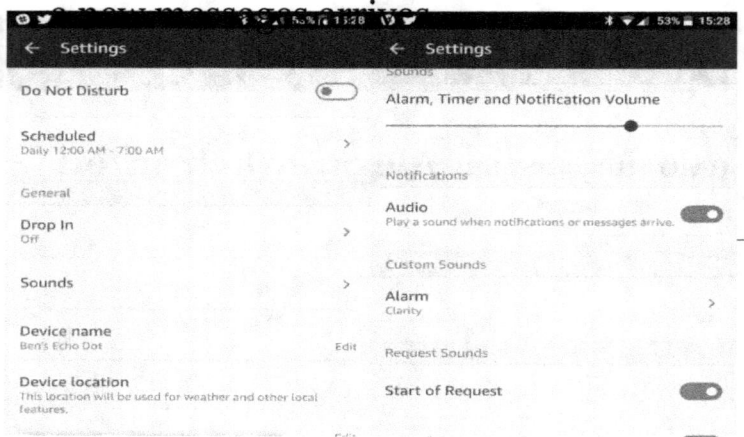

We recommend that everyone who owns an Echo switch on both the **Start of Request, and End of Request** sounds. Anytime you say **Alexa,** your Echo will play a little tone to let you know that she heard you. Same tone will be heard after your Echo recognize that you have finish speaking.

CHANGING YOUR ECHO DOT DEVICE LOCATION

Normally, your Echo should set its location automatically. But in case, just go to:

Settings

Device

Device location, and set your address. This helps you get the most accurate information when requesting about local details.

ACHIEVE SHIPPING NOTIFICATIONS ON ECHO DOT

If you wish, Alexa can notify you when your Amazon packages are close to delivery. Go to:

Settings

Notifications

Shopping Notifications

And switch on the **Shipment Notifications.**

Anytime you find a yellow ring, tell Alexa to read out your notifications to see when your item is arriving

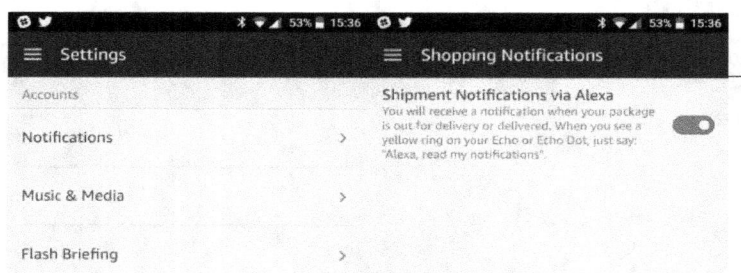

EDITTING ECHO DOT NEWS SOURCES

Once you are here, you should go to **Settings,** and then **Flash Briefing** section to change where your news is gotten from. By default, your news source is set to NPR's hourly news summary and the weather. But if you wish to add sources, just select **Get more Flash Briefing content** and include any you like. But we may suggest **MakeUseOf's Tech News skill?**

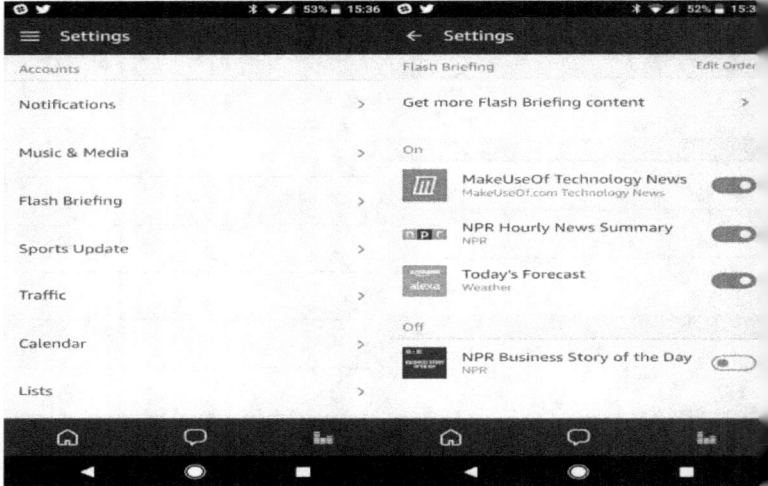

INCLUDING BEST SPORTS TEAMS TO ECHO DOT

To do this say **Alexa, sports update** and she will tell how your best team are doing and their next game. But firstly, you should be specific about which team you support. To get this done head to **Settings**

Sport Update menu.

To add team to your list, just search for a team name and hit it to include it into your list.

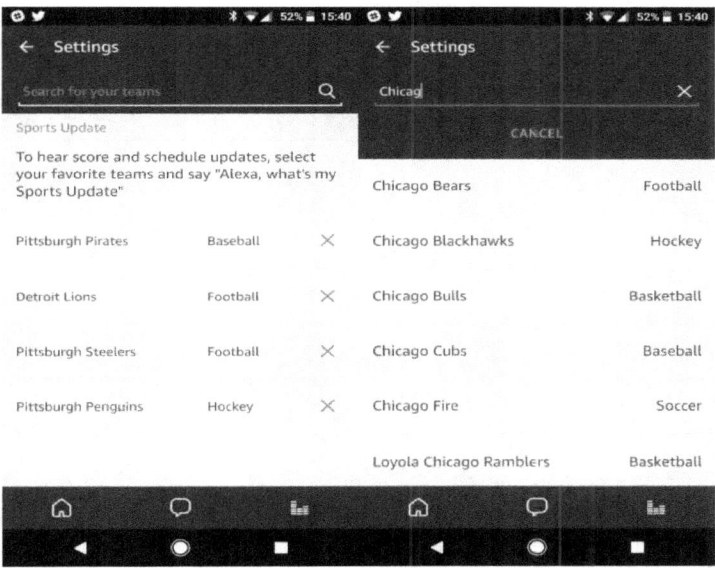

MAKE COMMUTE ON ECHO DOT SPECIFIC

Recall when we said Alexa's traffic finding ability? Move to **Settings,** and then **Traffic** to be specific about your daily commute. Begin with your home address, then specify the destination of your work. You can even include a stop in between if you always grab a morning coffee, or you drop your kids off school.

GET CALENDARS CONNECTED TO YOUR ECHO DOT

Alexa can include things to your calendar or inform you what's coming up in your day. To do this, first connect a calendar service. Head to **Settings**

Calendar

You can decide to connect your Google, iCloud calendars, or Outlook. Hit your preference and sign into your account to have them linked.

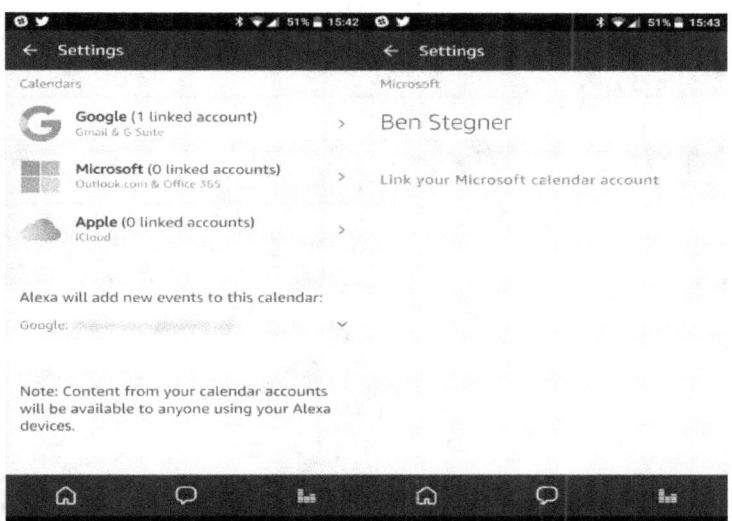

SYNC TO-DO LIST WITH YOUR ECHO DOT

The Alexa app add a basic to-do list. But if you already use another service, you would like to integrate it into your existing workflow. This can be done in:

Settings

Lists

Choose from the several popular services, Any.do and To-Do list inclusive, and then you can sign into your accounts and get your lists linked.

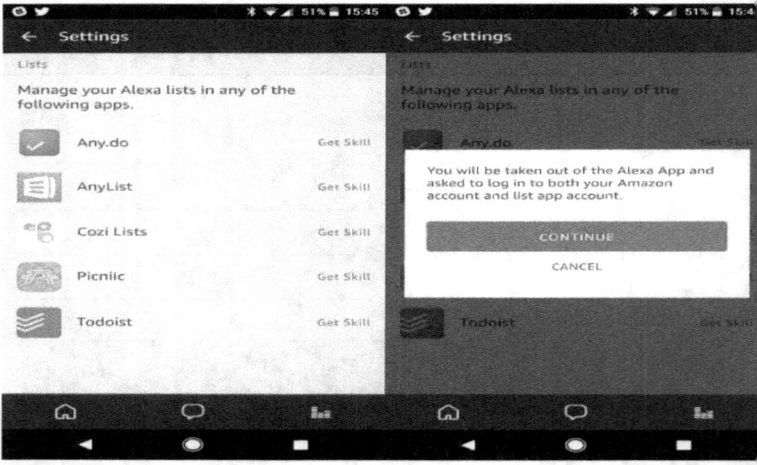

TRAIN ALEXA

If Alexa is unable to hear your request right, you can run through a fast training session. To train Alexa head to **Settings,** and then to **Voice Training.** After which, you should try to read 25 phrases in your normal voice while maintaining a distance. By using this method Alexa will be able to recognize how you sound.

ADD A PIN OR DISABLE VOICE PURCHASING ON ECHO

Using only your voice spending money on Amazon might sound too tempting for some people. If wouldn't want the skill to buy with Alexa, move to:

Settings

Voice Purchasing

Then disable the slider for **Purchase by voice.** Meanwhile, you can instead specify a four-digit code that is required during voice purchases. This serves as a prevention of guests from buying some trash and making a mistake.

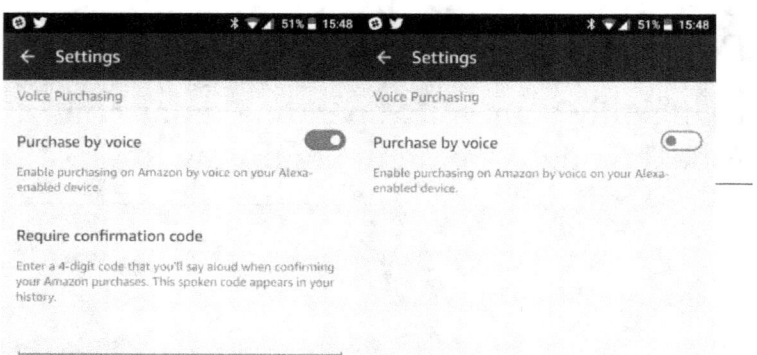

ENABLE MULTIPLE HOUSEHOLD MEMBERS ON YOUR ECHO DOT

Navigate to **Settings,** then **Household Profile,** arriving there you can add another user to your Amazon household. This allows both users access the other content and allows you share lists among other features. Amazon Household feature has uses even when outside of Alexa, so it's worth a try.

CHAPTER 7

TROUBLESHOOTING ECHO DOT COMMON ISSUES

Now that you are familiar with almost everything about your Echo Dot, let's finish up by talking about some common issues. In case any pops up for you, you should know how to troubleshoot it.

Note that, as rebooting your computer fixes many problems, the first and foremost troubleshooting step is rebooting your Echo. To power cycle your Dot, just unplug it, and wait a few seconds, then plug it again and allow it boot up.

ALEXA UNABLE TO HEAR YOU

When it appears Alexa is unable to hear you, try moving your Echo Dot away from any obstacles. It should be moved at least 9 inches away from any obstructions.

Also, try checking what other noise might make Alexa not to hear you clearly. You should know that an air conditioner working all day close to your Echo Dot will make it unable to pick up your voice, or even playing music with volume too high could hinder Alexa's hearing.

WIFI CONNECTION ISSUES

When setting up Echo Dot, it refuses to connect to Wi-Fi, or having it drop the connection randomly, Is a common issue. Having this problem, firstly try to power cycle all your networking gear, with the Dot, modem, and router. If still the problem stays, try moving your router and Echo close together. Also keep your Dot away from other devices, like microwaves that might interfere with it and hinder it from connecting. Or even disconnect other device on your network that you are not using to avoid waste of bandwidth.

ALEXA DON'T UNDERSTANDS YOU

Getting tired of asking Alexa a question and receiving a opposite answer instead. If Alexa doesn't hear well, you will get all kinds of weird reply. To remedy this is running through the voice training session we discuss above.

Navigate to **Settings**

Voice Training

Then Echo will tell you to speak 25 phrases from an accurate distance. This helps the device to recognize how you speak.

If still the issue still stunk around, you can check what Alexa feels you are saying. To do this just head to **Settings,** next to **History** in there you will find all your command you commanded Alexa. Hit an entry and you can play the live audio back, as well as confirming the Alexa did as you commanded. When addressing your Echo try to identify problem words it helps speak more clearly.

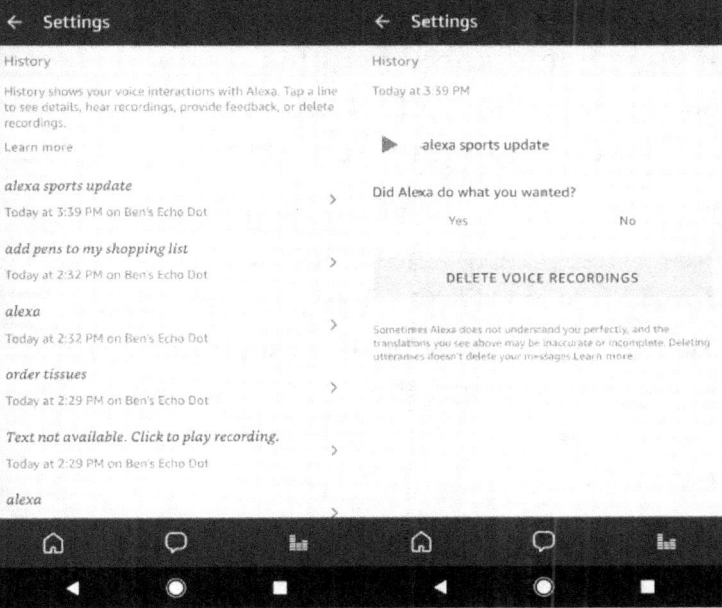

TOTALLY FROZEN?

FACTORY RESET YOUR ECHO DOT

After trying all the steps, we have outline above and your device still doesn't respond, you might need to take the nuclear solution. You can do a factory reset to send it back to default settings. But you should know that this step will erase all skills and preferences, so it will have to be setup again from start.

For the second generation Echo Dot, to factory reset it tap and hold the **Microphone off** and **Volume down** button together for about 15 seconds. After this you should find the light ring change orange, then blue. This indicate your Echo is ready to set up again.

The first generation Dot has a program reset button. Search for the small **Reset** button at the top of your device, then use a paper clip to hit and hold the button. The light ring will change orange, then later on blue and finally its back to factory defaults.

ALEXA, YOU ARE AWESOME

Now you have it, the complete guide to setting up your Echo Dot. You are ready for a world full of fun using this device. Whether it's just for weather, games, or even building an entire smart home around it, you still can't defeat what the Dot offers and the amazing stuff in it.

THE END